INTERNATIONAL POULTRY LIBRARY

JERSEY GIANT POULTRY

POULTRY BOOKS by Dr Joseph Batty

The Ancona Fowl Andalusian Fowl
Artificial Incubation & Rearing
Bantams -- A Concise Guide
Bantams & Small Poultry; Brahma & Cochin Poultry
Breeds of Poultry & Their Characteristics
Call Ducks: Concise Poultry Colour Guide
Cockfighting Ban of Oliver Cromwell
Domesticated Ducks & Geese
Duck Breed Books -- Aylesbury, Orpington, Indian Runners, etc.
Frizzle Fowl ; Garden Poultry Keeping
Hamburgh Poultry Breeds International Poultry Standards
Japanese Bantams: Japanese Long Tailed Fowl
Jersey Giant Poultry
Khaki Campbell Ducks & the Campbells of Uley
Keeping Jungle Fowl
Langshan Fowls
The Malay Fowl: Marans
Marsh Daisy Fowl Minorca Fowl
Natural Incubation & Rearing Norfolk Grey Poultry
Natural Poultry Keeping: New Hampshire Red Poultry
Old English Game Bantams
Old English Game Colour Guide
The Orloff Fowl
Orpington Fowl (with Will Burdett)
Ostrich Farming: Polish Poultry Breeds
Plymouth Rock Poultry
Poultry Ailments: Practical Poultry Keeping
Poultry Characteristics—Tails
Poultry Colour Guide
Poultry Foods & Feeding
Poultry for Beginners Poultry Breeding & Rearing
Poultry Shows & Showing
Poultry Table Birds
Races of Domestic Poultry, Sir Edward Brown
Revised Edition Joseph Batty
Rhode Island Red Fowl
Rosecomb Bantams
Scottish Poultry Breeds
Sebright Bantams
Sicilian Poultry Breeds
The Silkie Fowl
Sussex & Dorking Fowl
True Bantams
Understanding Modern Game (with James Bleazard)
Understanding Indian Game (with Ken Hawkey)
Understanding Old English Game
Welsummer Fowl Wyandotte Poultry

JERSEY GIANT POULTRY

DR. JOSEPH BATTY
Chairman:
World Bantam & Poultry Society Ltd

BEECH PUBLISHING HOUSE
STATION YARD
ELSTED MARSH
MIDHURST
WEST SUSSEX GU29 0JT

© BPH, 2006

This book is copyright and may not be reproduced or copied in any way without the express permission of the publishers in writing.

ISBN 1-85736-479-1
First Published 2006

British Library Cataloguing-in-Publication Data
A catalogue record for this book is available from the British Library.

Beech Publishing House
Station Yard
Elsted Marsh
MIDHURST
West Sussex GU29 0JT

CONTENTS

1. Development of a Giant   7
2. Jersey Giant Standards   27
3. Jersey Giant Colour Standards   43
4. Notes on Management   49
   INDEX   79

AUTHOR'S PREFACE

The Jersey Giant is going through a revival, although too slowly for a breed with such excellent qualities. It is very hardy, lays quite well and is an excellent table bird. Moreover, the hatching and rearing come up to a high standard, the chicks growing quite rapidly.

It is therefore a fine utility breed which has many desirable features. However, if shown, size alone is not enough. The breed is not expected to be exceptionally large and coarse -- if size alone is wanted why not go in for turkey breeding ?

Despite its size, there is no loss of symmetry and beauty and this applies without losing its utility characteristics.

In Britain, bantams are quite scarce, yet they have great possibilities; they are a good size at over 3 lb. which makes them suitable as table birds for those which have to be culled !

My thanks are given to the many artists and authors, past and present, who made my research more fruitful. And Good Luck to those who aspire to take up this fine breed and increase the numbers. !

Joseph Batty   November , 2005

FREE RANGE JERSEY GIANTS

JERSEY GIANTS

1

DEVELOPMENT OF A GIANT

They are big, strong and very hardy birds, and the ideal life for them is on free range, roaming over the fields at will.
Margaret Arthur, an early enthusiast.

Black Pair & White Hen Jersey Giants
Based on a painting by H Hoyle

DEVELOPED FOR A MARKET NEED

There are ample records to show that the Jersey Giant breed came about as a natural evolution from farmers in New Jersey, USA who aimed to produce a large fowl which catered for the needs of the local market. Although the process appeared haphazard the objectives were known by the farmers of the Burlington County area who sought to satisfy their customers*.

These objectives were:

1. Large fowl with plenty of breast meat and ample thighs.
2. Black was preferred.
3. Skin to be yellow.
4. They had to be profitable.
5. Meat rather than eggs would be the aim, although this would involve keeping the birds up to about 11 months of age for mature capons to be produced. (about 14 lb).

Later, attempts were made to ensure that the breed laid a reasonable number of eggs and, as a result, a total of 200 eggs was being achieved.** However, not all agree with the wish to make them top egg producers because this might interfere with the objective of producing top table birds.

The aim therefore was to produce a utility fowl which fattened naturally, no doubt running around the farmyard and in a nearby field. Often, as was usual in those days, the poultry would be a side line, being looked after by the farmers' wives.

When ready, the birds would be crated and sent off to market, usually to cater for a festive day.

---

* The Jersey Black Giant, Willard C Thompson, in *The Feathered World*, 4th Sept. 1925.
** The Jersey Giant Fowl, Louis A Stahmer, in *The Feathered World*, 14th August, 1931.

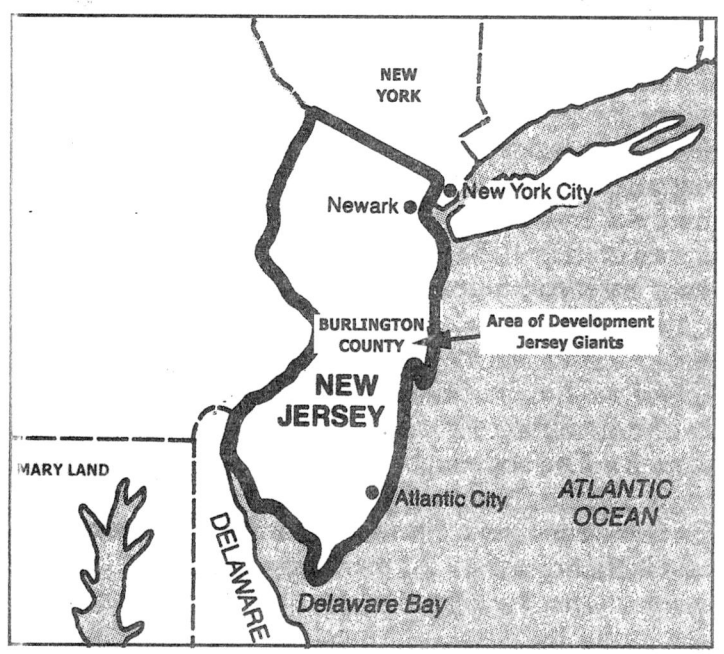

Area of Development

The original development of the Black Jersey Giants was in Burlington County, New Jersey.

BREEDS INVOLVED

The breeds involved in the development were:
1. Farmyard poultry
2. Black Javas
3. Dark Brahmas
4. Black Langshans

and later

5. Cornish (Indian Game)

The belief is that the original concept of providing table birds for the large markets in New York, Philadelphia, and the Atlantic Coast commenced in the 1870s possibly by two brothers named John and Thomas Black who were farmers in Burlington County, but whether this is entirely true is not clear, because the breed was named Jersey Black Giant which appears to refer to the colour. Later this was modified to Black Jersey Giants and these were accepted into the American Standard of Perfection in 1922, which meant that more than a quarter of a century elapsed before the breed became 'official' in the USA. Later, a White was added and admitted to ASP in 1947. Brown also suggests that Partridge Cochin and Barred Plymouth Rock were brought into the development.* Since the latter were around in the 1870s this cross is a distinct possibility because it was one of the most popular breeds.

This is the accepted version of the development of the breed as appearing in the ASP and many other places. However, an alternative explanation suggests that the large black breed; in effect, the double of the Black Jersey Giant came from Spain.**

---

* *Poultry Breeding & Production*, Edward Brown, London, 1929.
** *Poultry Breeding & Management*, Wm. W Broomhead, London, nd

The records purport that Christopher T Turle, then Secretary of the Jersey Black Club in the UK (1927) was told the story of the breed being found in Gibralter in 1885-7 and birds were purchased from **Algeciras** which is a town in Spain near to Gibraltar. This fact which might be taken as indicating a Meditteranean origin as the Jersey Giants do not come broody very much. (Broomhead, *ibid*).

This story is not authenticated in any other place although it may not have come to light until after 1929 when Brown had his findings published. Even so, no modern writer appears to take this alternative source. Other factors are:

1. The eggs are light brown, known as *tinted* in the UK, but Spanish breeds lay white eggs.

2. There appear to be no large breeds in Spain.

3. The skin and flesh of the Mediterranean breeds is white and the egg is white.

Whether these alleged birds from Spain ever reached the USA is not clear. Moreover, there seems little doubt that the breeds named above were used to produce the brown egg and yellow skin. Acordingly, even if the story has some truth in it, there is little doubt that the breed which came out as the Black Jersey Giant was one which had been greatly modified to meet the needs of the market.

There has been a Pea Comb variety although this is not a standard type and this may have been the result of the cross with Cornish (Indian Game) mentioned.* They never became popular in this form.

---

* *The Jersey Giant Ideal*, Louis A Stahmer, (ibid)

DEVELOPMENT BREEDS EXPLAINED

The breeds, which were involved in the development, are now explained:

1. Farmyard poultry
2. Black Javas
3. Dark Brahmas
4. Black Langshans
   and, later,
5. Cornish (Indian Game)
6. Partridge Cochins
7. Barred Plymouth Rocks

FARMYARD POULTRY

Which of the farmyard poultry were involved cannot be stated with certainty. There are no acknowledged native breeds in the USA so any breed which was around from early days could be the type; obviously, it would preferably be on the large side.

In Britain we usually take Bewick's Fowl as being the typical farmyard type, although in England the Old English Game and Dorking were present at the time of the Roman invasion more than 2,000 years ago. A drawing of Bewick's Fowl is given overleaf, demonstrating the typical common breed, somewhat resembling the early type of Dorking before it was spoilt by unwise crosses to make it better for showing. A Malay cross was involved at one stage.

In the USA the Dominique and Black Java appear to have been present from early times, but certainly not more than a few hundred years. No doubt birds were taken from Britain and others were imported from other countries before the period when poultry breeding became very popular around 1840.

BLACK JAVA FOWL

The Black Java fowl was developed in the USA, but how it arrived in that country is not clear. Claims were made that it originated in Missouri, but this appears to be a case of carrying patriotism too far. If it was an American native fowl how did it get there? Moreover, in appearance, there is clear visual evidence that the Java has its origins in Asia. Whether the breed was related to the Langshan with its hard, glossy feathering is not known, but this seems likely, which would also bring the brown egg gene.

Following years of controversy the American Poultry Society state in their *recent* Standards that the breed originated in the Far East, although the breed has been modified since that time. The English approach appears to be acceptance of the Asian origins, although the exact country is still a mystery.

> **Opposite Black Javas**
> At first claimed to be an American breed, but now acknowledged to be of Asian origin,

Bewick's Fowl (1800)

Thomas Bewick (1753 - 1828) was an engraver noted for his accuracy.

The Dark Brahma

The Dark Brahma is given as one of the main breeds for developing the Jersey Giant and this would before the extra plumage and tall, upright body were developed fully. The breed was involved in developing many others, including, notably, the Barred Plymouth Rock.

Later, catering for exaggerated show features, the Dark Brahma grew larger, more feathery on body and legs and very stately and regal in appearance. The changes which took place can be seen from the two illustrations opposite. The drawing by J W Ludlow would depict the Victorian ideal. If the date of the development of the breed is taken as after 1870 it would mean that the top illustration, showing the primitive form of Brahma, would have not been the type taken, but, rather, something approaching the Ludlow ideal, but remembering that the illustration was of an English type of Brahma, whereas the American variety would have been a little different.

The existence of feathering on the shanks and feet on the Brahma could have caused a problem, although, as it transpired, this was not serious and was fairly quickly bred out.

Early Form of Dark Brahma

Dark Bramas Modified for Showing (Ludlow)

BLACK LANGSHANS

The Croad Langshan is large in body, fairly long and wide, without the massiveness met with in the Cochin, Brahma and Plymouth Rock; breast deep and long, carried well forward; back rather long and sloping, with the tail rising sharply therefrom; the head small in comparison with the size of body, carried well back, and full over the eye; beak light to dark horn, the latter for preference-- comb medium in size, single, carried upright in both sexes, evenly serrated, and fine in texture, which, with the face, earlobes and wattles - the latter small and a brilliant red; eyes large, bright and sparkling, brown to dark hazel, with black pupil; neck long and well arched covered fully with long hackles; wings medium in size, carried well up; tail full, with abundant sickles, thus differing entirely from the Cochin, and carried high; legs medium in length, standing well apart well feathered on the thighs, but with no signs of vulture hocks, and the shanks and outer toes slightly feathered; toes, four in number on each foot, long and straight. The bone should be fine; the shape of the bird is graceful, sprightly and active, with an appearance of nervouness; feathers carried close to the body, and brilliant in sheen.

Weight: Males 8 to 10 lbs.; females, 6 to 8 lbs.

The '*Croad* Langshan' is so named because this was the original type imported into England by a Major Croad. Later, there were modificatications, including a tall breed known as the *Modern Langshan* a purely show bird. The American version is in between the Croad and the Modern, although there is no suggestion that the Americans crossed with the Modern Langshan.

Undoubtedly it is an important breed which carried the dark brown egg gene which was passed on to other breeds.

Original Type of Langshan

There are differences between the Langshan and the Jersey Giant, such as a short back and wrong type of tail, discussed in the chapter on the standard.

CORNISH (INDIAN GAME)

This is very much a 'developed' breed containing a mixture of blood from Old English Game and Aseel which, through careful selection has resulted in a broad breasted breed which matures slowly.

They were developed by Cornish miners who used them for cockfighting. In Britain they are called *Indian Game*, but the Americans decided that in their country they should be named Cornish Game.

As a table bird they have great merit, fattening quite naturally. At the time of the cross for Jersey Giants they would be longer on the leg than the modern Indian Game and would possess greater vigour.

Old Style Indian Game

PARTRIDGE COCHIN*

Sir Edward Brown described the Cochin in the following terms:

> In general appearance the Cochin indicates a large, strong bodied fowl, broad, deep and full in front, and thickly covered with feathers. Massiveness is characteristic of this fowl. The back is short and broad, rising to the tail, which is small and round, and free from long sickles; the head is small and neat in comparison with the proportions of the body; the comb is single, small in size, standing perfectly upright, and evenly serrated; the wattles long and fine and the ear lobes are long and bright red; the legs and thighs are short, widely set apart, and thickly covered with feathers, as are the hocks, the latter often curled around the joints; the bone is very heavy, and the entire appearance indicates strength.

The Partridge variety was introduced in 1874 just about the time of the development of the Jersey Giants. The description is as follows:

MALE: Black Red type of bird with black breast and tail and remainder golden red, with some variation in shade. The hackle is striped at the base with black, free from gold shafting. The saddle and wing bows are a burgundy red. Wing bar black and wing bays a reddy brown. The leg and foot feathering is black, but may include some gold.

FEMALE: Partridge markings, with an overall golden brown colour. The neck should be marked with black stripes, within the gold, but not showing the shafts. The tail is a dark colour (blackish and gold) and the body is evenly covered with a golden brown with pencilling on each feather.

---
* See *Brahma & Cochin Poultry*, Joseph Batty, BPH

 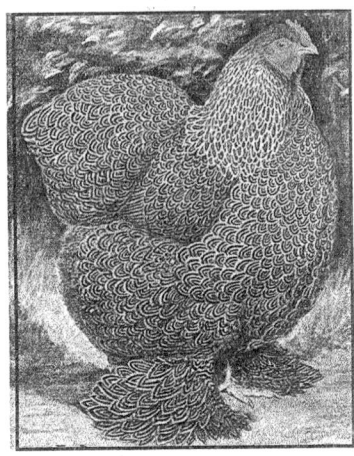

Partridge Cochins -- Male & Female

As noted these are a form of Black Red but the female has fine lacing on each body feather.

The reason for the selection of Partridge Cochins is unclear because they had nothing to offer in terms of colour, and they have feathered shanks which are not required for Jersey Giants. There was the size factor and this may have been the reason they were used.

BARRED PLYMOUTH ROCK*

Plymouth Rocks are an early American breed which possessed many utility features. In appearance they are quite handsome, being heavily built yet well balanced. They are of medium length in body and quite deep, with full breast which yields ample meat. They live and produce under any reasonable conditions possessing great vigour and stamina. Whether on free range or on limited runs they still thrive. They yield large-sized brown eggs and some strains have been known to lay as many as 250 eggs in their first full year of production. Mature birds reach around 9.50 lb. for cocks and 8 lb for hens (4.39 and 3.63 k respectively). They fatten well so broilers (table birds) can be produced in 8 - 10 weeks. In addition, they are quite mature at 8 months old or even a little earlier, and this is at the normal rate.

In blood lines they are similar to the early Jersey Giants the breeds involved in their development being Javas, Cochins and other Asian breeds with the addition of the Dominique which introduced the dark barring on a light background.

The introduction of this breed was no doubt due to its popularity in the USA, being large and possessing great vigour.

---

* See *Plymouth Rock Poultry*, Joseph Batty, BPH

Very Early Barred Plymouth Rock

Dominique Hen
This breed was used to develop Barred Plymouth Rocks. Barring is black with an inhibitor which causes the barring, so on colour it was a good choiice.

WHITE JERSEY GIANT

As noted, the White Jersey Giant came on the scene possibly in the 1920s, being a 'Sport' from the Blacks. There are many references to the variety from 1931 onwards.* They were being bred in Britain from that date a well known breeder being James Cowan. Also a Mr and Mrs Turle had them on free range, also importing a cock from the USA.

Why they were not standardized in the USA until 1947 is not clear. They have the great advantage of white plumage which is much better than black for a table bird.

There were also claims that the skin was white, but this is not specified in the Standards (ASP), although in Britain white skin would be preferred for any table bird.

Since the shanks are willow in colour, getting a perfectly pinky white skin might be difficult, but not impossible.

White Jersey Giants on Free Range (1930s)

---

* For example the article by Stahmer (ibid) and *The Feathered World Year Book*, 1934. An article by James Cowan appeared in that journal.

BLUE JERSEY GIANT

Getting a Blue colour would not be difficult from the Black and White. There has been a suggestion that Orpingtons were at some stage introduced into some strains so crossing with a Blue Orpington would be a feasible method of getting the colour.

This colour is listed in the British Poultry Standards, but not in the American (ASP). It is suggested that the blue should be laced, but why is not clear, because breeding Blues is difficult anyway.

Multiplicity of Colours

The Jersey Giant is a rare breed so with its increasing popularity attention should be paid to the Blacks and Whites.

BANTAMS

Getting bantams for the very large breeds has always been difficult. British Standards lists them at weights of just over 3 lb (1.74 k.) for cocks, and 2.50 lb (1.13 k.) for hens.

In the ASP the Whites and Blacks are listed as 36 oz and 34 oz. for cock and hen respectively. They appear to be quite scarce, even rare.

STANDARD FOR LARGE IN BRITAIN

The large fowl Black Jersey Giants were admitted to the British standards on 14th March, 1924. There is no date for the Whites being admitted to the standards, but they were certainly around before they were standardized in the USA.

---

* James Cowan, ibid.

2

JERSEY GIANT STANDARDS

The standards are the written Ideals for each breed.

Jersey Giant Cock

Must be deep in body with full, broad breast and stout thighs to give an excellent table bird.

The hen is similar to the cock except for the sexual differences and size (see overleaf).

JERSEY GIANT POULTRY

DETAILED REQUIREMENTS

1. BODY.

Rounded in shape but within a squarish area but not fully globular like the Wyandotte. This should include:

(a) BACK: Longish back which is broad and a body which is deep.

(b) BREAST: Broad, curved breast. Carriage should allow breast to be in line with the end of the beak.

Faults in Jersey Giants

Must be sure that breed standards are applied and not the standards of other breeds. Compare the Ideal opposite.

Black Jersey Giant Hen Ideal

See also the features of the male bird two pages back.
Both drawings were done by American artist A O Schilling to illustrate the ideals.

The labels have been added by the author of this book to high light the main features.

JERSEY GIANT POULTRY

Faulty Cock
This bird looks too much like the Plymouth Rock

Faults
Jersey Male
Short Body

Fault in Hen
Deep Body
Too Rounded

(c) TAIL rising from the the back, but without excessive cushion. Medium, rounded in the male and at an angle of 45 degrees for males and 30 for females. It should be neat with the feathers overlapping and quite broad. There should *not* be the high mound (cushion) of the Cochin.

The tail of the male has rounded feathers which form into an almost circular shape.

Tail Measure
Male 45°

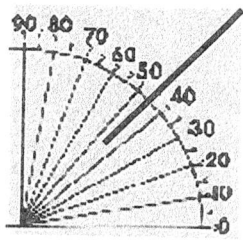

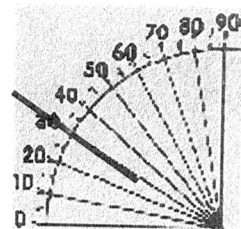

Tail Measure
Female 30°

Angle of Tails Above Horizontal
Examples of faulty tails are shown on the next page.

Jersey Giant Hens With Tail Faults

Jersey Hen With Baggy Back End
May be a prolapse or indicating too fat.

2. LEGS set at a slight angle and located just back from the midway point of the body.

These should be set apart, supporting the body, yet allowing a bird to be agile and nimble on its feet. There should be no hint of bow-leggedness or turning inwards.

The feet should be strong with four toes which should not be twisted or thick at the joints. The scales should be regular and gleaming, with no scars or irregular marks and no sign of scaly leg (caused by a parasite).

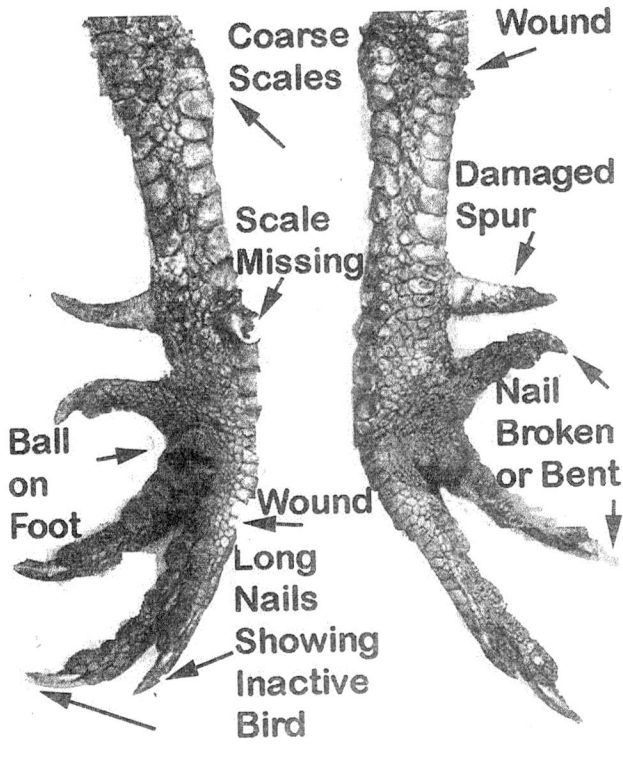

Faulty Legs

FAULTY LEGS

These apply to all breeds and are a very serious fault.

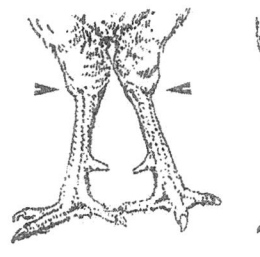 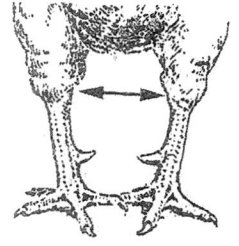

Knock Knees   Bow-Legged

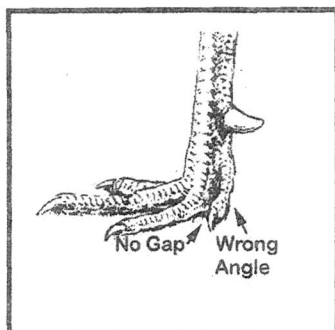

Duck-footed
A Major fault.

Web Footed

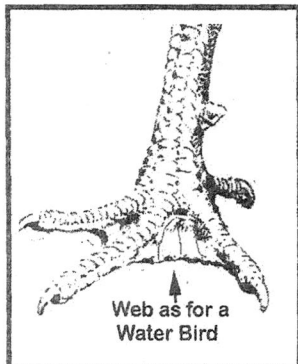

*Drawings after
John H Robinson*

Jersey Hen With Shanks Too Long

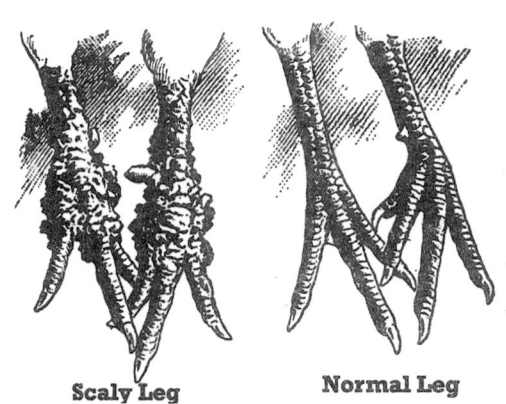

The Dreaded Scaly Leg Disease
Wash shanks and rub with antiseptic cream
if disease suspected.

JERSEY GIANT POULTRY

3. **WINGS** to be medium in size (relative to size of breed) and carried horizontally across the body, with no signs of looseness or split in the middle or twisted or broken feathers. The point of the wing should meet the base of the tail and be hidden by the saddle hackle.

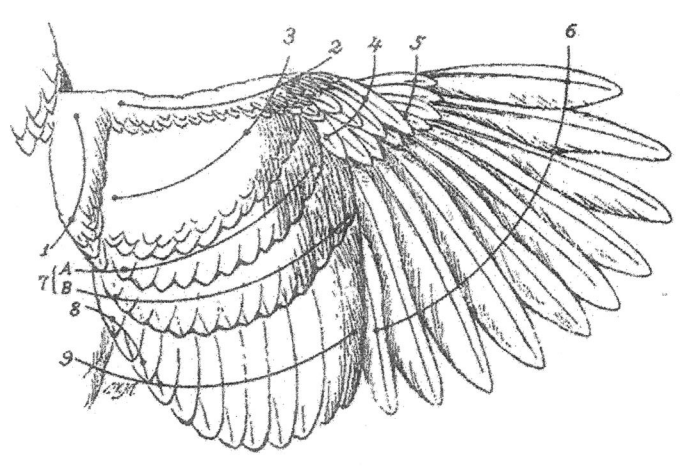

1. Shoulder
2. Wing Front
3. Wing Bow
4. & 5. Primary Coverts
6. Primaries (Flights)
7. Wing Covers (Wing Bar)
8. Tertiaries
9. Secondaries (Wing Bay)

Essential Features of the Wing
Missing and broken feathers will be penalized when showing.

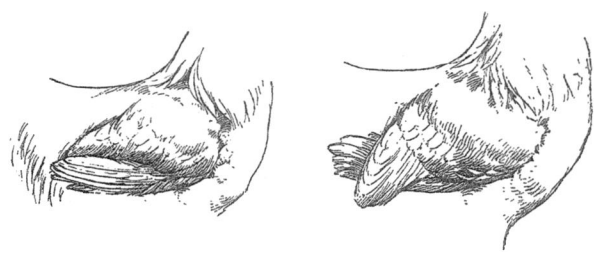

Faults on Wings

Jersey Giant Pullet
Winner, but not a good specimen. Head and neck weak. Thighs inadequate. Face much too white and comb faulty.

4. **HEAD** Tending towards being round, but flattish on top, although this cannot be called flat in the sense of a straight line. However, it should be well balanced and not appear stunted in any way -- neatness is essential. The face should not be coarse.

Neck: Longish, broad at base, arched, and well covered with hackle feathers, without too much fluff. The hackle should not be twisted or appear untidy.

The neck is well arched and this gives the impression of flatness, whereas it is the natural curve of the head and neck.

Comb: This is medium in size and upright showing six spikes well spaced with the middle one higher than the rest. The comb of the female smaller. Faults should be penalized.

Wattles & Ear Lobes: These should be oval and medium in size, with the cock's slighter longer, and the ear lobes should be oblong in shape reaching down to about one-third the length of the wattles, angled across the face. They should be a sparkling red in colour.

A fault is *round* ear lobes, especially if these are white in colour.

Eyes: These should be quite full and bright. There should be no beetling of the brows. The colour should be dark brown although slightly lighter allowed in Whites. Light eyes are faults.

COMB TOO LARGE. TAIL TOO SMALL

Jersey Male With Faults

Comb Upright Follows Contour of Neck

Neck Longish Arched. Medium Feathering

Wattles Neat, Longish & Rounded at Ends

Beak Shortish Slight Curve. Black shading to Yellow in Colour.

Eyes Bold & Large -- Brown Ear Lobes Oval & Sloping. Fine, Red.

Head of Male Black Jersey Giant

POSSIBLE FAULTS

The main faults generally found are as follows, but CONDITION is also vital as well as conforming to TYPE; ie, in accordance with Breed Standards, but above all correct SIZE. Small birds to be penalized or even passed.

1. Wrong Colour Shanks; eg, yellow.

2. Ear Lobes; marked, not even, white colouring.

3. Wattles short or round, disfigured.

4. Feathering on Legs; scaly legs; discoloured.

5. Twisted Toes; other than four toes.

6. Comb: faulty or marked; wrong type; large and coarse; too many spikes or other fault.

7. Face Colour: not bright red; white patches or blemishes.

8. Legs: knock kneed or cow hocked; cocks without spurs; coarse with dark marks; duck footed; stilty.

9. Tail: wrong size or angle; cocks lacking curved sickles; broken tail feathers.

10. Wrong coloured plumage, white in Blacks calls for disqualification. In Whites or Blues wrong colours.

11. General: any other faults which are not in accordance with the *Standards*. For example: white skin (except in Whites, but this is questionable); beetle brows, wrong colour eyes; faulty beak; short neck; faulty wings; lacking green lustre in black parts; twisted or curled hackle feathers; breast narrow with pronounced keel.

Cup Winning Monstrosity

This strange cockerel won 1st and Cup at Olympia in 1924. Fortunately, the breed has been greatly improved since those very early days. It does not resemble the standard in any way.

3

JERSEY GIANT
COLOUR
STANDARDS

Both are good types for those early days.

Black Jersey Giants in Britain in 1930s

MAIN COLOURS

The main colours are:
1. **Black**
2. **White**
3. **Blue**

The Black was the original colour and remains the most popular.

BLACK VARIETY

The Black variety should have black plumage with a positive green sheen. The under colour should be a dark grey, some latitude being permitted, but certainly not a very light colour. The American Standards specify *dull black* under colour for black birds with dark legs..

There should be a shine or lustre on the plumage with the male being more prominent.

LEGS: Black or near black with the sole of the foot yellow.

COMB and FACE: Bright red with no white colouring or blemishes.

EYES: Dark; brown ideally, but a slightly lighter colour may be permitted, but certainly not pearl or very light.

BEAK: Blackish with yellow showing through at the tip.

Serious Faults in Colour

Any white on the plumage or other colour which is not black. Legs should be as black as possible, but there is a tendency to run to grey or willow which could be acceptable if dark.

WHITE VARIETY

The White variety should have white plumage with a positive sheen. The under colour should be white or, at least, very light.

There should be a shine or lustre on the plumage with the male gloss being more prominent, especially on the hackle.

LEGS: Willow although many do have legs which are greyish, which appears to be acceptable. Getting the correct shade in willow can be difficult because it borders on grey and a muted green. Yellow legs are a disqualification.

COMB and FACE: Bright red with no white colouring or blemishes.

EYES: Dark; brown ideally, but a slightly lighter colour may be permitted, but certainly not pearl or very light. The early breeders preferred an ebony black eye because this sets off the white bird much better than a light eye.

BEAK: Horn colour with a yellowish base. Again, some latitude is permitted with some whites having a dark horn colour.

NOTE: When white chicks are bred and hatched they turn out a better white if the original under colour is grey. These tend to give better colour adults than chicks which are white when hatched. (James Cowans, *ibid*).

BLUE VARIETY

This colour is quite rare although there should be no difficulty in breeding it. If laced, as indicated in the British Standards, there would be more difficulty in achieving a high standard because Blues are difficult any way, and the addition of lacing even more so.

The British Standards state the Modern Langshan Blue should be followed and this is along the following lines:*

Blues are a slatey-blue, with the male bird showing a darker head and neck hackles. The American standard shows lacing on all the body feathers, including the tail of the hen; the cock does not have the lacing on the tail or hackles. In general, the colour follows the Blue Andalusian fowl, and this is stressed for the German Langshan. In Britain there is not the same insistence on lacing for Blues, but since the colour is non-standard, and rare, this fact is really academic.

Possible Faults -- Colour

The main faults arise from trying to achieve a medium blue colour. Possibilities are:

1. Wrong colours such as red or black in plumage.
2. Eyes, legs, and beak the wrong colours which should be similar to the Black.
3. Lacking lustre on feathers.

---

* See *The Langshan Fowl*, Joseph Batty, BPH

POINTS TO PONDER

The Americans claimed this was the largest breed of poultry and, undoutedly, they can grow very large, but based on the standard size, it should be appreciated that the Dorking cock at maximum is listed as 14 lb. which is one pound heavier than the Giant. However, some well over standard weights have been recorded.

On the standard, both American and British, there are a few anomalies.

The colour of the legs is not altogether clear. Black is stipulated for the Black variety but illustrations show dark grey. With the Whites the legs are stipulated as Willow, but the painting by Hoyle shows greyish legs. (see Colour illustration shown earlier)

Under-colour appears uncertain with one prominent breeder stating that the best chicks should have grey under colour, but this does not comply with the ASP.

In Britain the preferred colour of the skin is white and this was being claimed by some breeders in this country. However, this is against the American standard; also, with willow legs, there is more likelihood of getting a yellow or off white skin than a pure white.

These are relatively small points on what is a very worthwhile breed which develops naturally into a large fowl. The chicks are hardy and a joy to rear. More attention should be paid to developing the breed on a much wider scale.

4
NOTES ON MANAGEMENT

POULTRY MANAGEMENT

Nest Box for Broody and (below) the **Correct Positioning of the Eggs and formation of Basin.**

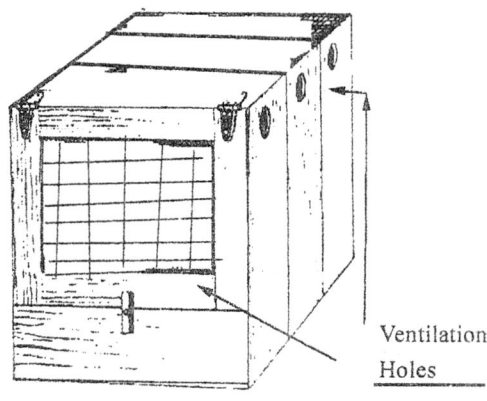

Ventilation Holes

Gauze or Small Mesh Wire netting

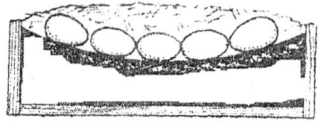

1. Correct Positioning

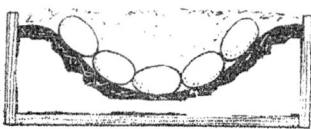

2. Incorrect Positioning (Too Deep)

HATCHING AND REARING

USING BROODY HENS

For small numbers the broody hen rules supreme, and the best results will usually be obtained by this method. A Silkie cross is usually the most suitable.

USING AN INCUBATOR

There are two or three small incubators on the market which will produce satisfactory results with all eggs. The "visual" type is quite popular and these have a perspex top through which the eggs can be observed. However, other types may be better insulated. Take care to see what you are buying, because no amount of automatic controls can make up for strength and durability; those made of weak plastic should be avoided.

The instructions issued with the particular incubator should be followed closely. Experience suggests the following hints should be followed:

1. Use fresh eggs not more than 7 days old. Mark them with the date in pencil, but use a felt pen to show the date when eggs are put into the incubator.

2. Eggs should be a normal, oval shape with strong even shells (porous shells are not usually hatchable).

3. Locate the incubator in a room which is well insulated and free from sunshine or draughts (fluctuations in room temperatures are better avoided).

4. Keep the water tray topped up, so that the eggs have adequate moisture.

5. Turn the eggs at regular intervals, usually twice per day.

6. When chicks start to hatch try to avoid opening the incubator.

7. Check the thermometer at regular intervals so that the incubator keeps steady at around 103°F (39.4°C). Exact temperature depends on type of machine and maker's recommendation.

8. Load the machine with a reasonable number of eggs; e.g. at least 20, or so few chicks will be hatched that rearing will be uneconomical.

Success with an incubator requires care, and application of the rules issued with the machine. Checking for fertility is advisable at 7 and 14 days. Any clear eggs should be removed. With experience it becomes possible to use a strong torch or special Candling Machine, and detect fertile eggs which have become "addled". However, in the early stages the beginner is advised to remove the "clears" only.

BROODIES

Once the chicks have hatched they can be transferred to a broody hen—one which has been sitting for at least two weeks. Alternatively, a brooder can be used; this may simply be an infra-red lamp placed in a suitable shed with a canopy or metal shade.

In hatching under hens you need not confine your attention solely to bantams as broodies; but it is *not* wise to use heavy hens of 7 or 8 lbs. weight, or you will suffer broken eggs and crushed chicks. If you must use large hens, try medium-weight utility birds such as Buff Rocks, which sit lightly on the eggs and are careful mothers.

Smaller broodies are, of course, better and usually the bantam breeder gets enough sitters amongst his own flock; but those who require stocks of broody hens might consider keeping a small flock of Silkie-crosses. These are better than pure Silkies, which have several **disadvantages.** They desire to lay and sit again somewhat quickly, whereas many chicks usually require an extended rearing period; they are very susceptible to scaly-leg, which is transmitted to the chicks; and their silky plumage sometimes wraps itself round the necks of chicks and strangles them. Nevertheless, some breeders swear by them and would have no others.

POULTRY MANAGEMENT 53

Silkie Crosses: Make Excellent Broodies & Mothers

Silkies on Tethers Being Exercised and Fed.

After a period (10 minutes) make sure they return to the correct nest and close up. For safety the hatching boxes are best under cover in a large shed or barn.

As noted, the best broody hen you can get is a Silkie crossed with a clean-legged variety such as the Wyandotte. This is a great favourite with large-breed fanciers and has even been produced commercially. Their "staying-power" greatly exceeds that of **pure Silkies**.

Whatever you do, avoid scaly-legs in broodies, or it is absolutely certain to attack your chicks. Even hens that have been treated are not good—the paraffin or other cure makes them restless, and they sometimes quit before hatching.

For bantams your hen is best not heavier than about 2 kilos and is therefore usually a bantam cross or pure breed. For large eggs use a heavy breed which is noted as a broody type. The light breeds such as Leghorns are generally unsuitable and, in any case, are not likely to come broody anyway. Dust her with insect powder *well before* the eggs are due to hatch. Treat her with kindness. A roughly-handled broody gets nervous and may not readily settle down.

For broody hens, cheap nests like apple or orange boxes have much to offer, which can be burnt after use and so avoid vermin. Fresh nests for every batch of hens are excellent. About 14in square (11in for bantams) is the recommended size. It should be large enough for the nest and the hen, without too much extra space or corner spaces, which allow eggs to fall out of the nest proper and get chilled.

It is often recommended that the foundation of the nest should be earth, or a turf moulded saucer shape. The chief virtue of this is that it shapes the nest nicely; and if a damp piece of turf is used it provides moisture for the eggs. Have a retaining board at the front of the box.

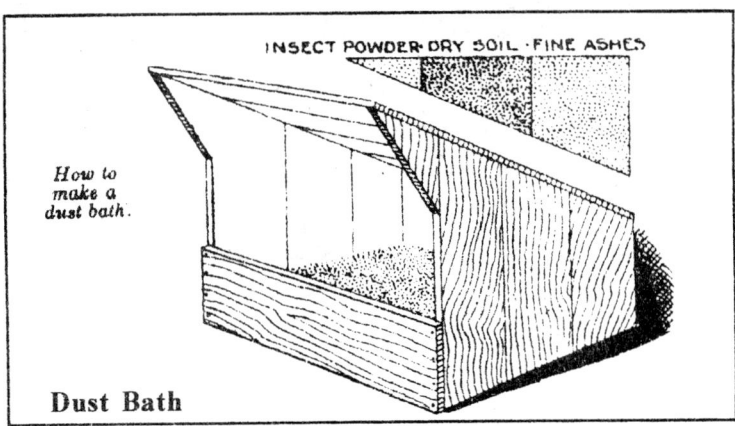

Dust Bath

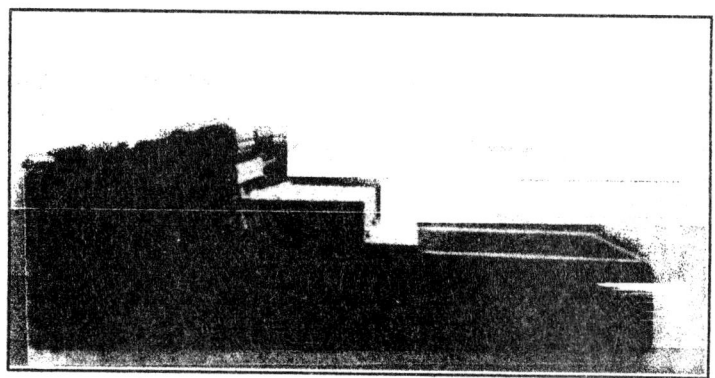

A Small Chick Coop and Run

A Group of Chicks Recently Hatched

Warmth is vital —these are huddled together which is a sure sign they are feeling chilled. Make sure there is a surround so they cannot stray from the infra-red lamp, especially the first few days when they are very vulnerable.

NEST OF SOFT STRAW

Whether or not you use earth as a base, the nest itself should be of soft straw. Hay is not so good, as it is dustier and more likely to attract vermin. Oat straw is best; make the nest saucer-shaped, put a few pot eggs in it, and show it to the hen. More often than not she will walk on and settle down—though if she is a stranger she may need coaxing and shutting on the nest. Some of the best are restless for the first few days.

When she returns to her nest of her own accord after about ten minutes' feeding and watering she may be trusted with the eggs she is to hatch. Give her these at night, when she is quiet and settled. Don't give her every egg she can cover, particularly in cold weather. Two fewer are better than one too many, or chilled eggs and a bad hatch will result.

Don't startle hens by sudden noises. Keep them quiet or you will experience trouble. Their vicinity is no place for hammering up new sitting boxes or repairing wooden fences. They are best fed on grain if possible, as soft food is likely to promote diarrhoea.

Nests fouled by broken eggs or excreta should be cleaned, the straw replaced, and eggs washed in warm water. This should not occur if the hen is comfortable, but it happens all the same. If the hen has diarrhoea, slip down her gullet a lump or two of gum arabic moistened in cod-liver-oil twice a day for two days. This usually effects a cure. If not, try feeding with bread soaked in milk, squeezed dry and dusted heavily with french chalk.*

Egg Showing Shrinkage of Air Space

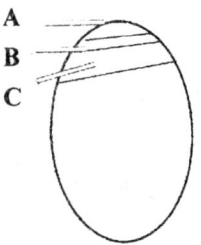

KEY

A = 3% at 8 days

B = 5% at 10 days

C = 10% at 17 days.

---

* These are old fashioned remedies—modern drugs should be effective.

POULTRY MANAGEMENT

Chicks Being Reared Under Broody Hens

These would be in a vermin proof enclosure with someone on duty to keep a close watch on them.

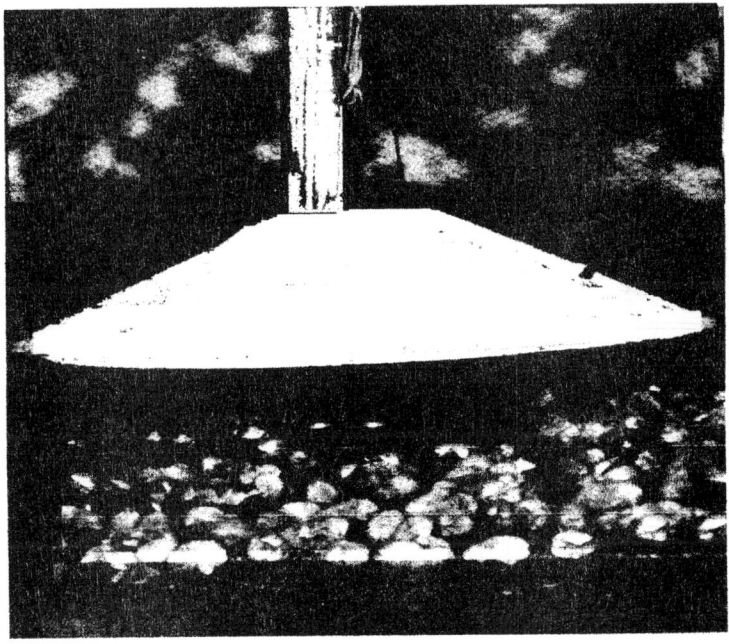

Rearing by Infra-red Lamp

FEEDING CHICKS

Feed the chicks from the second day with chick crumbs which contain a high level of protein (around 20 per cent). Do not try to economize on food because the chicks eat very little. Some chick crumbs contain a drug which combats coccidiosis, and this is recommended by some breeders, but others prefer to use drugs only if the chicks contract the disease on the grounds that this helps to build up resistance.

For rapid growth Turkey chick crumbs may be used, but are not always suitable, because if fed too long on a very high level of protein, birds mature very quickly.

Water should be in a small chick fount and changed daily. Broken mixed corn and chickweed can be introduced at about 3 weeks; in fine, warm weather they can be put out in a wire-netting covered run on the grass.

A full diet of Growers' Pellets can be fed from about 8-weeks of age, but make sure the chicks get plenty of greenstuff and titbits such as soaked bread, and, where appropriate, a limited amount of household scraps. For hard or medium feathered birds, such as Game, feed mixed corn as well, thus keeping the feathers tight and glossy. Spare cockerels, not to be kept on, can be fattened by introducing layers' pellets and wet mashes.

The broody hen may be fed with mixed corn and later with layers' pellets. She may suffer from loose bowels at first so clean any mess away or the chicks get soiled. Clean shavings should be put in the coop, but not too deep or the chick water fountain will get clogged up, and they will suffer from thirst which is quite serious for young chicks which must have clean, fresh water at regular intervals.

At four or five days, test fertility by holding eggs at night before a strong light. The best way is a box with a hole for the egg and an electric light inside. Fertile eggs will clearly show blood veins and the germ spot, the whole at the fourth day looking like a spider's web.*
With light shelled eggs you can test readily at the third day, but with dark brown eggs the seventh day is usually best.

---

* See *Artificial Incubation & Rearing*, J Batty, where the different stages are shown in colour.

POULTRY MANAGEMENT

The size of the air space is an important indicator to the state of the incubation process. With the natural loss in weight during incubation the content shrinks from about 3 per cent at 8 days to 12 per cent just before hatching.

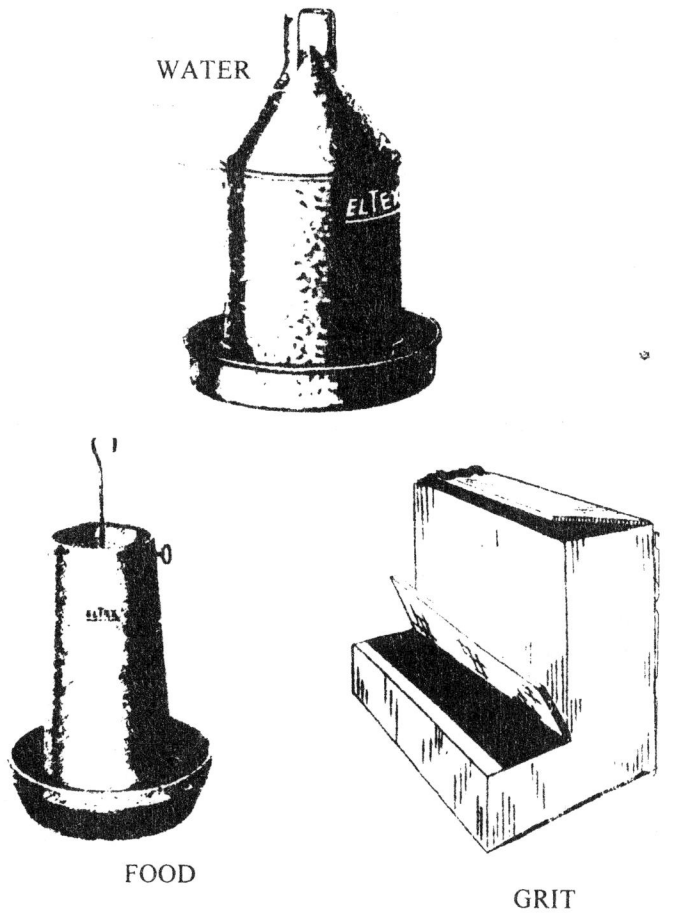

Basic Food & Drink Utensils

FEEDING

Feeding is relatively simple and depends on a number of factors:

1. Laying Birds

Adults may be fed layers' pellets and mixed corn. In winter, if the birds are not laying, mixed corn may be adequate, but change to layers' pellets of the small side when laying is about to commence— usually early in the new year. About 18 per cent protein is required for a laying bird or to condition a cock.

Greens such as chickweed and grass clippings should be made available as soon as growth starts in the garden or fields.

2. Chicks and Growers.

These require special chick crumbs to start with and they are gradually weaned over to Growers pellets and *broken* mixed corn, starting about 5-8 weeks of age, but making sure they are kept on the crumbs until they are quite well feathered.

If there is access to grass and scratching out of doors all the better, but do not put chicks out too early, especially in cold wet weather.

Ensure that food such as pellets is kept dry and this can be achieved by having hoppers with lids which are kept inside or in some form of shelter.

Water fountains are also essential, preferably galvanized so they do not rust. Moreover, they should be of the type which allow the top to be removed so that the inside can be cleaned thoroughly.

Greens' Rack

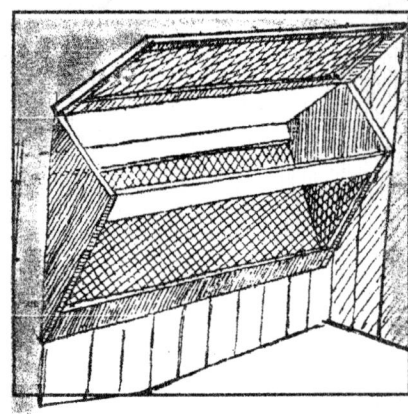

POULTRY MANAGEMENT

Special Shed Suitable for Bantams.

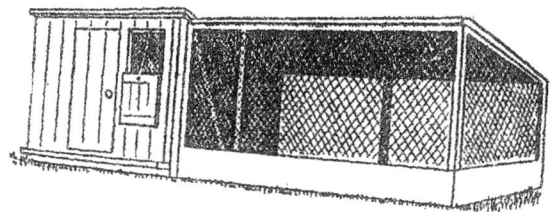

House & Open Fronted Scratching Area

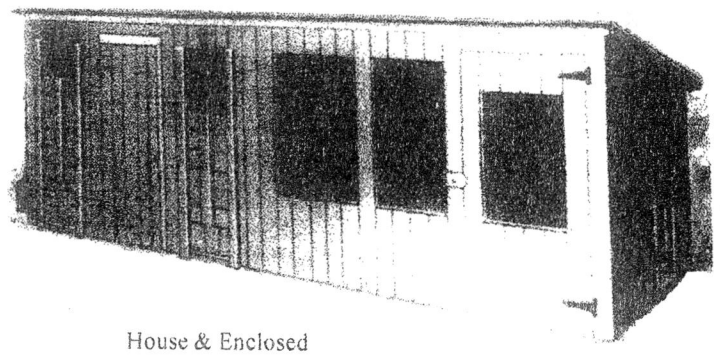

House & Enclosed Run.

Various Types of Shed for the Small Scale Breeder

SMALL SHEDS AND RUNS

Try to plan sheds so they can accommodate all necessary equipment and can be managed easily.

A wide variety of sheds and runs are available or can be made by a poultry keeper with a reasonable flair for carpentry. Many designs are possible to fitt the space available

Possibilities are:

1. **House and Run Combined.**
2. **House and separate Run —this should be covered over or closed off in winter.**
3. **Sussex Ark or similar movable 'house'.**

There has been recognition for many years that for winter eggs light must be provided and the shed, whilst adequately ventilated, should also be well insulated so that the birds feel warm enough to lay. The lighting can be arranged on a time switch or simply by turning on the light each evening at dusk and then turning it off when a total of 12 - 15 hours light has been given, ie ; make up the difference with artlflclal light.

In designing a hen house it is essential to watch out on *economies of management*. Keep a dustbin or similar container with lid, inside each shed, to store layers' pellets, so feeding is easy. Make sure there is provision for greens by provldlng some form of **greens' rack**.

A dust bath should also be available, filled with sand, cinders and fine earth, which the hen scatters on her to kill mite. For those with limited space or for winter quarters a garden shed is ideal. A shed 3m. X 2m. can be divided into two compartments with a door between and two trios can be kept, one in each section. However, be sure to have hardboard along the bottom of the partition because cocks *like to have a*

POULTRY MANAGEMENT

go and therefore they are better separated. Large breeds will require twice as much space. Also include a platform shelf about 2 ft. wide across the back and about 2- 5 ft from the ground, the heavy birds should be kept low. This acts as a droppings board and a nest box can also be included. The height gives the birds exercise as they fly up to reach the food and perches.

The floor should really be concrete or very strong wood (tarred) to keep out rats and mice. This should then be covered with garden soil, or shavings and leaves and other materials added to form deep fitter for scratching. The fancier who desires to keep many birds will require a large shed with a corridor up the middle and compartments on each side with separate doors. Trap doors will then lead out into separate runs so there is ample space for breeding pens. In addition, if showing, or when spare cockerels are kept, or there is a health problem, a Penning Room or shed will be essential for training and/or getting birds into condition.

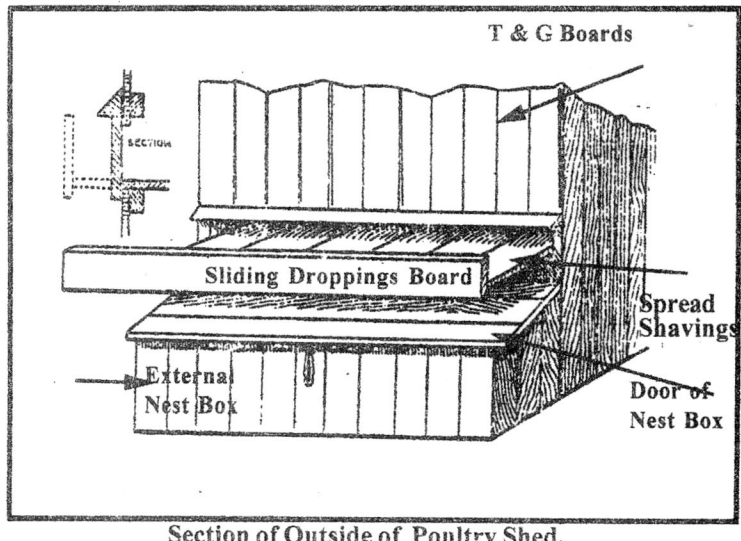

Section of Outside of Poultry Shed.
The Droppings Board can be pulled out for cleaning.
Perches are above the Board.

IMPORTANT FACTORS

A number of factors should be considered when designing or purchasing a suitable shed. These are:

1. Space

Generally the more space the better, but having regard to the economics of cost, a shed measuring 2 metres x 1.5m would house about 6 layers and if a separate, large shelf (a mezzanine floor) is erected the birds will fly up and it gives them exercise as well as extra space.

2. Run

Ideally the birds should have a run which has a grass base because grass is a fine food. An orchard is ideal because the hens can run around and scratch. Alternatively, a Sussex Ark or similar shed which is moved once a week can provide fresh ground. The latter is essential because birds kept too long in one place develop disease from the faeces of any hens which have health problems. To combat this problem, when space is limited (in a 'fixed' run), the poultry keeper must be constantly building up the surface with soil and clippings and once a year it should be limed and turned over — preferably being rested at that time. Soil and grit are essential to poultry so the addition of fresh material as turves or other form of earth is ideal.

3. Dry and Comfortable

Birds must be kept dry, free from draughts and damp. This applies especially in winter and therefore make sure the shed is water proof, has shutters to keep out the rain and cold and is adequately lighted.

Internally it can be insulated with hard board, but make sure the mice cannot get in the space between the wall and the insulating board. A very good material is *Styrafoam*, but this is expensive. A foil-coated back to the insulating board and some form of cavity lining is obviously very good for retaining the heat.

4. Position

The house should face south and should be on sandy soil. No hens can continue to lay and be healthy on muddy land. Put pebbles on the outside around the 'pop-hole' so that mud is not carried inside.

5. Cleaning

Have a small hand shovel and bucket so that all faeces can be removed on a regular basis —preferably daily, thus avoiding any danger of soiling the eggs. put sand, earth or shavings on the droppings board and have litter on the floor at least 3 inches thick (7.50cm).

6. Egg Collection

Daily collection is vital and clean eggs must be the aim. Put in a refrigerator or, if for hatching, put in a suitable cool place and turn daily.

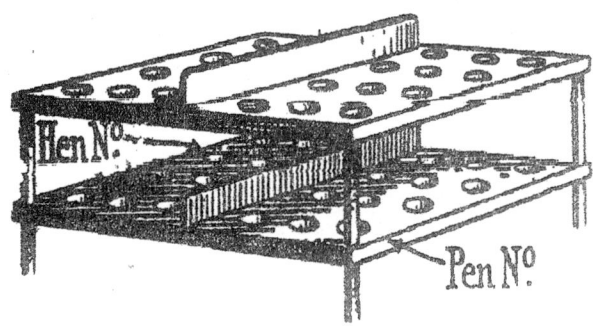

Special Egg Tray for Showing Source of Eggs
The egg or a label would show the essential details
of date and Pen.

PENNING & TRAINING ROOM

Anyone keeping a reasonable number of poultry should aim to have a separate room in which wire cages or wooden cages with wire fronts in which birds can be rested or trained for shows. This should contain a food bin which is vermin proof, exhibition baskets, first aid cabinet and, if at all possible, electric lights and power and a sink with running water.

Irrespective of the colour of the bird it will benefit from being washed, a process which also keeps down the lice and mite. Birds presented for shows should be checked very thoroughly, especially round the vent which becomes red and sore from lice or mite when birds are neglected.

The Standards of Perfection

Inevitably with the shows there had to be guide lines to indicate how birds had to be judged. Thus were the first *standards* issued, in Britain in 1864, for the first Poultry Club which foundered, because its ruling body could not agree. The standards were then modified and agreed at the Poultry Convention held in New York in 1872 and these were finally adopted in Britain in 1886.

The process was aided by personalities of the period; Lewis Wright, W B Tegetmeier, and W F Entwisle all played a part and ideals were drawn by Ludlow, thus showing what was was expected.

Harrison Weir also provided illustrations for the earliest of the poultry classics, but his birds were drawn as they were, realistic and practical, rather than the Ludlow-type fowls of beauty which probably never existed except in his imagination.

MODERN SHOWS

After the vigour and enthusiasm of the Victorians, with the many new breeds being developed, it was inevitable that there should be a period of refinement and consolidation. Not that new developments ceased; new breeds were still being created, although many fell by the wayside. The World wars also affected the situation, as did controls imposed to control fowl pest.

Today there are still many shows at local, regional and national levels; in Britain they are controlled primarily by the Poultry Club, but there are many shows put on by breed clubs, as well as the major national show organized by the Federation of Breed Clubs. The Rare Poultry Breeds Society also plays an important part in supporting breeds which are in small numbers and do not have a breed club.

REQUIREMENTS

The procedure for shows has changed very little over the 150 years of their operation. They are planned, schedules sent out, entries made by exhibitors, penned on the day, judged, awards made and then collected by the exhibitors and returned back to their homes.

In summary form the requirements are as follows:

1. Breed or purchase birds which comply with the *STANDARDS OF PERFECTION.*
2. Feed and house the birds correctly.
3. Train them and ensure they are in top condition.
4. Take them to the Show in peak condition and washed.
5. Take only the best birds.

The dedicated fancier tends to frown upon the practice of buying in for showing. It must be appreciated that a winner is only a winner in the right hands: many winners are purchased only to find that they no longer win because they fail in condition or other requirement.

POULTRY MANAGEMENT

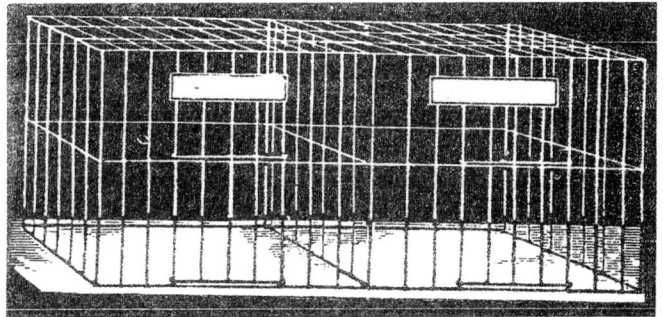

Typical Show Pens
Similar cages should be used for training the birds.

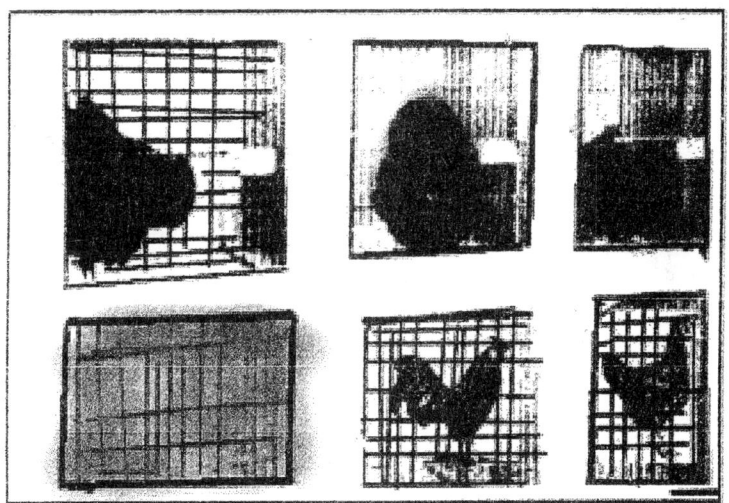

Section of Training Pens in Penning Room or Shed.
These are made of wood with wire fronts and a number of rows can be arranged. Keep the water containers clean and replenish daily.
Handle the birds regularly to train for shows and also use a judging stick.

FEED & HOUSE THE BIRDS

If showing is to be taken seriously, special special attention should be paid to the show requirements. A fancier cannot seriously expect birds which have been running around in, say, a breeding pen, to be caught up and taken to the show and hope to win. There must be a Penning Shed or room containing show cages or similar wire fronted boxes which allows the birds to be kept indoors, fed with tit bits and special food, the occasional tonic and regular handling so that they are quite tame.

Even the main houses are best when they are the size of garden sheds which allow the fancier to stand inside and 'talk to the occupants', thus developing an informal and friendly atmosphere so the birds are not afraid. Lifting birds from the perch or from the training pens when they have gone to roost allows the condition to be assessed.

Some breeders have cockerel pens which consist of a small shed and run and these are kept outside. They are excellent for some breeds, but provide extra shelter in bad weather. They may be used for a breeding pair in summer or for running cockerels or for a broody hen and her chicks. If they are outside in a paddock or on a lawn the birds may be frightened by cats or foxes, and they will not allow a bird to be trained for the shows, but generally, for a person with enough space, they are very useful.

The Penning room may also be used for a hospital room, although any birds with infections should be kept away from other birds and if there is any serious disease the sick birds may be better killed.

Do not keep birds in limited accommodation for long periods; otherwise they become too fat and suffer from *dry hackle* which, in turn, leads to mite in the feathers. In warm weather, spray with water and ensure there is adequate light and ventilation.

POULTRY MANAGEMENT

Handle the birds regularly and go through the *judging process*, which teaches them to be unafraid at shows.

BLOCK OF CAGES

SINK

CORN BIN

CUP-BOARD

BENCH

REQUIREMENTS
Baskets
Bowls
Medical Box
Shelves
Dubbing Scissors
Ointments
Olive Oil
Vaseline

BLOCK OF CAGES

PLAN OF PENNING ROOM
Rows of pens at eye level, plus electric light.
May be modified as necessary.

Plan of A Penning Room

Train the Birds & Make Sure They are Fit

The show pens allow birds to be trained so they are tame. They can be handled regularly and examined to make sure the feathering is correct. The author prefers wooden cages about a metre long and about 0.75 metres wide with wire fronts fitted with access doors — one at each side.

The bottom should have a good covering of soft wood (white) shavings which absorb all the moisture and this should be changed regularly, or at least the soiled material removed, and more shavings added. A wooden block or a perch may be added to allow the bird to perch.

Water in a metal cup hanging on the side should be changed daily and food should be given on the same basis, possibly pellets and tit bits in the morning and mixed corn in the evening; remove any surplus food each day and try to assess the exact needs so the birds are well fed yet remain fairly hungry. The occasional mixture of cod liver oil on the corn or pellets and a few drops of *Abidec* or similar vitamins are also recommended by some fanciers, but this is not essential provided a balanced mixture of food is given, as well as chick weed or other greenstuffs. Grit is also essential and a charcoal preparation for birds is said to improve the digestive sytem.

Using a Judging Stick

A judging stick may be used to train birds to stand in a certain way and to appear to the best advantage.

Try to train them to "pose" to best advantage, so they stand with breast curving out, with head high and the tail of the cock formed into the correct position. Moreover, if trained properly the backs look the correct length; even hens benefit.

POULTRY MANAGEMENT

As noted earlier, keeping birds in show cages can improve condition and add on plumpness, but they may become too fat and lethargic. Nails grow long and have to be clipped, but watch out to avoid cutting into the blood vessels.

Because of these factors try to change birds around and, in the summer, let every bird have a period in the sunshine; it is unkind to do otherwise. In the poultry fancy most fanciers frown on battery cages so why have prize birds in close confinement. At all times each bird must have adequate room to exercise and, remember, no cock is likely to be fertile after a period of very close confinement and this applies very much to the true bantam which has problems unless the correct conditions and food are provided.

If inbreeding is practised which is inevitable to some extent, because outcrossing often loses the essential show features which have been 'fixed' and therefore locked in, fertility may also be affected.

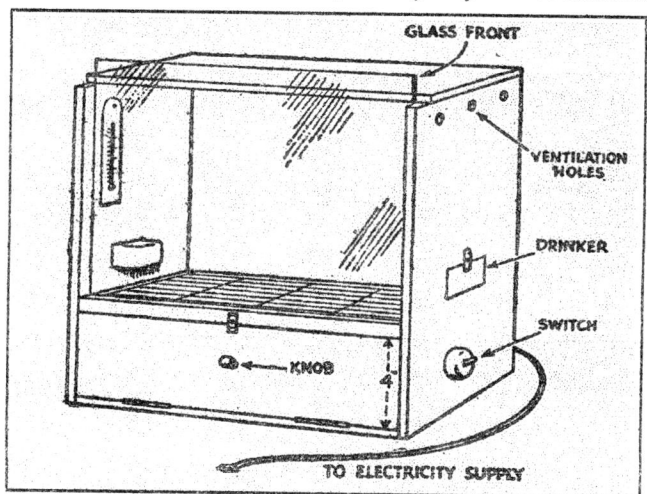

A **Drying Cage**, which should be large enough for a bird to turn around. Two 40w bulbs or an infra-red heater should be sufficient to achieve the heat required. Do not overheat and if possible have different levels of temperature.

WASHING

The need to wash birds must be acknowledged because they should never win if dirty or in poor condition. In any event, washing reveals any defects that may exist and are not apparent from a superficial examination. The points to watch are as follows:

1. Wash about 4 days before the show so the feathers are fully dry before exposing a bird outside and in the show. Feathers which are not dry never hang properly and therefore the bird in question is at a disadvantage.
2. Use warm water at a temperature which allows the hands to be immersed, yet warm enough to fetch off the dirt.
3. Most fanciers use two bowls, some three, to wash thoroughly and then rinse in one or more lots of clean water at a temperature which is just warm. The first bowl should contain washing-up liquid or a hair shampoo, and later bowls are for rinsing and adding any conditioner. The cleanser must not be too strong or it will take too much of the natural oils from the feathers.
4. Wash the legs first, outside the main bowl — usually on the side of a sink so the dirt can be washed away and not go in the washing bowl. Use a medium stiff scrubbing brush to remove all the dirt.

Drying

The birds should be dried in a warm room, but not too hot. Various methods are used, including a drying cabinet rather like a hospital cage with built-in heater, a small hair dryer, or in front of a fire or radiator in clean, straw-filled baskets.

After drying the birds are put back in the show pens and kept as clean as possible and immediately before the show the legs are washed again and polished with a cloth. Remember no faking as such is allowed, although it must be appreciated that the birds must be turned out to the best advantage.

POULTRY MANAGEMENT

HEALTH

Never show birds with health problems because under any decent judge they will be 'passed' so they cannot appear in the prize list. Sitting in a mopy position, eyes running, scaly legs, faeces on the feathers near the vent (and mite), feathers with tiny holes in them, barely visible, but signs of feather mite, and a multitude of other obvious faults, will all show the judge the state of health.

The judge must handle each bird and check for condition as well as any defects. The correct colouring on the plumage, including any defective feathers, should be established by physical examination; in this way there can be no quarreling with decisions made. A judge who has seen a defect can speak with authority, but if he adjudicates only from *outside* the show pen there are likely to be mistakes.

Minor Problems, But Signs of Mismanagement

To avoid mite, cleanliness must be observed. Moreover, dusting powder should be used and the occasional clearing of feathers from around the vent followed by an application of a a general ointment or vaseline. When washing for shows a little disinfectant in the water will also keep away feather mite, but avoid allowing this to get in the eyes.

A Judge at Work
James Barr, an expert, judging a show.
Handling birds is essential or many faults will be missed.

A more serious problem is *scaly leg* caused by a small parasite that gets under the scales and disfigures the bird. The scales become raised and uneven and scurfy. The birds are unfit for shows and they become carriers of the disease. They should not be shown until cured.

As a boy, the author remembers his father using a mixture called 'Troopers Ointment' which was very effective, but is no longer available because of poison regulations. Soaking and washing the feet in hot water containing disinfectant and a liquid soap can be the start of the cure. After that the legs are rubbed thoroughly with sulphur ointment, if available, or Benzyl Benzoate a liquid. *The only way to avoid the problem is strict cleanliness of the legs and feet, washing regularly (say, once per month) with warm water, containing washing-up liquid and a mild disinfectant.*

Remember the legs can make all the difference to the appearance of a show bird; one with dirt ingrained, with long nails, should be penalised by the judge. Neglected birds, not cleaned, may even smell in an unpleasant way. and that calls for down marking irrespective of any merits which a bird may possess.

Final Touches

Before taking birds to a show and when penning put the finishing touches by wiping over the face, comb, and legs with white spirit and wipe the lobes with baby powder rubbed gently on to the wet lobe, which should help to whiten them. This may be done a number of times. Then remove the powder before penning.

Some fanciers use a special mixture for a head and legs dressing.to be used sparingly. One recommended by the late H Easom Smith was:

2 oz. spirits of wine. 1 oz. **camphorated oil.**
1oz. glycerine. $1/_4$ oz.. **citric acid.**

NOTES ON BREED

INDEX

A
Ailments 75
Air Space 56

B
Barr James 76
Breed 1-48
Broody Hens 52, 53

C
Candling 58
Chick Rearing 55, 57
Coop 55

D
Droppings Board 63
Drying Cage 73, 74
Dust Bath 54

E
Eggs 66
Entwisle W F 67

F
Feeding 58-70
Food 58

G
Greens 60

H
Handling 65
Hatching 51
Health 75

I
Incubation 51

J
Judging 72

L
Ludlow J W (Artist) 68

M
Management 49 - 78
Mite 75

N
Nest 50, 56
Nest Box 50

P
Penning Room 67, 70, 71
Poultry Club 68

R
Rearing 60

S
Scaly Legs 75, 77
Sheds 61, 62, 64
Show Pens 69
Show Preparation 77
Shows 68, 77
Silkies 53
Standards 67

T
Tegetmeier W B
   (Poultry Pioneer) 67
Training Birds 69, 72
Training Room 67

U
Utensils 59

W
Washing 74
Weir, Harrison 67
  (famous poultry artist)
Wright, Lewis
  (Author) 67
Wyandotte 54